GUIDELINES FOR DEFINING PROCESS SAFETY COMPETENCY REQUIREMENTS

GUIDELINES FOR DEFINING PROCESS SAFETY COMPETENCY REQUIREMENTS

Center for Chemical Process Safety
New York, NY

Published by John Wiley & Sons, Inc., Hoboken, New Jersey.
Published simultaneously in Canada.

For general information on our other products and services please contact our Customer Care Department within the United States at (800) 762-2974, outside the United States at (317) 572-3993 or fax (317) 572-4002.

Wiley also publishes its books in a variety of electronic formats. Some content that appears in print, however, may not be available in electronic formats. For more information about Wiley products, visit our web site at www.wiley.com.

Library of Congress Cataloging-in-Publication Data:

Guidelines for defining process safety competency requirements / Center for Chemical Process Safety, New York, NY.
    pages cm
  Includes index.
  ISBN 978-1-118-79522-4 (hardback)
  1. Chemical processes—Safety measures. 2. Manufacturing processes—Safety measures. I. American Institute of Chemical Engineers. Center for Chemical Process Safety.
  TP150.S24.G854 2015
  660'.28—dc23   2015005680

Printed in the United States of America.

10 9 8 7 6 5 4 3 2

DEDICATION

This book is dedicated to Scott Berger, the 4th Executive Director of the CCPS. Scott's vision of a global process safety community led to the first Global Congress on Process Safety in 2005, which continues to grow in participation and now supports regional process safety conferences in Latin America, Asia and the Middle East. In Scott's 13+ years as Executive Director, the membership in CCPS has expanded to over 180 member companies representing 27 countries. During Scott's tenure CCPS has continued to mature and is a recognized source of sound, unbiased technical expertise in process safety which is valued worldwide. For that, the process safety community owes Scott recognition and our sincere gratitude.

CONTENTS

List of Tables   *xiii*
Files on the Web Accompanying This Book   *xv*
Acronyms and Abbreviations   *xvii*
Glossary   *xix*
Acknowledgments   *xxi*
Preface   *xxiii*
Executive Summary   *xxv*
Organization of This Book   *xxv*

1. INTRODUCTION   **1**
  1.1 Why Process Safety Competency?   1
  1.2 Purpose   2
    1.2.1 Address Applicable Regulations and Requirements   2
    1.2.2 Identify Process Safety Competency Requirements   3
    1.2.3 Assess Individuals Against Recommended Competencies 3
  1.3 Audience   3
  1.4 How to Use This Process   4
  1.5 Risk Based Process Safety Elements   5
    1.5.1 Process Safety Competency   6
    1.5.2 Corporate Process Safety Culture   8
    1.5.3 Process Safety Knowledge Management   9
    1.5.4 Organizational Change Management (OCM)   10
  1.6 CCPS Vision 20/20   10
  1.7 References   11
2. IDENTIFY PROCESS SAFETY ROLES & COMPETENCY NEEDS   **13**
  2.1 List of Generic Job Roles   13
    2.1.1 Facility Manager   14
    2.1.2 HSE Manager   15
    2.1.3 Operations Manager   15
    2.1.4 Inspection, Testing and Maintenance Manager   15
    2.1.5 Supervisor   16
    2.1.6 Operator   16

2.1.7 Technician   16
2.1.8 Engineer   17
2.1.9 Project Manager   17
2.1.10 Project Engineer   17
2.1.11 Additional Roles   18
2.2 List of Proficiency Levels   18
2.2.1 Awareness Proficiency Level (Level 1)   18
2.2.2 Basic Knowledge Proficiency Level (Level 2)   19
2.2.3 Practitioner Proficiency Level (Level 3)   20
2.2.4 Expert Proficiency Level (Level 4)   20
2.2.5 Leader Proficiency Level (Level 5)   21
2.3 List of Process Safety Knowledge/Skills   21
3. PROCESS SAFETY COMPETENCY MATRIX   **23**
3.1 What Is the Matrix?   23
3.2 How to Customize the Matrix   25
3.2.1 The Proficiency Levels Worksheet   25
3.2.2 The Risk Based Process Safety Worksheet   25
3.2.3 The Skills and Knowledge Worksheet   30
3.3 Uses of the Matrix   31
3.3.1 Establishing a Training Matrix   31
3.3.2 Organizational Changes   37
3.4 References   38
4. INDIVIDUAL AND ORGANIZATIONAL PROCESS SAFETY
COMPETENCIES   **39**
4.1 Develop Organization Specific Competencies   39
4.2 Assure Compliance with Regulations   41
4.3 Example Templates and Checklists   44
4.3.1 Process Hazards Management Coordinator and Hazard
      Assessment Facilitator   44
4.3.2 HAZOP Facilitator   45
4.4 References   46
5. ASSESS COMPETENCIES VS. NEEDS   **49**
5.1 Assessing Existing Competencies   49
5.1.1 Self-Assessment   50
5.1.2 Peer/Manager Assessment   50
5.2 Training for Assessors   50
5.3 Identify Gaps Between Current Status and Needs   51
6. DEVELOP GAP CLOSURE PLAN   **53**
6.1 Methods for Closing the Gaps   54
6.1.1 Tasks or Personnel Reassignment   54
6.1.2 Internal & External Training   54
6.1.3 External Resources   55
6.2 Supporting Materials   55
6.3 Pre-requisites Before Progressing to the Next Level   57
6.4 Example of Managing Gap Closure   57

7. SUSTAINING COMPETENCIES   **59**
 7.1 Strategies for Sustaining Competencies   59
  7.1.1 Apply Changing Technology & Lessons Learned   59
  7.1.2 Metrics   60
  7.1.3 Demonstrate and Verify Competencies Are in Place and Effective   61
 7.2 Review and Update Competency Needs   61
 7.3 Organizational Process Safety Culture   62
 7.4 References   62
APPENDIX 1: EXAMPLE COMPETENCIES FOR AUDITING   **63**
APPENDIX 2: PHM COORDINATOR & HA FACILITATOR
      QUALIFICATIONS   **65**
 A2.1 Purpose   65
 A2.2 Assumptions   65
  A2.2.1 HazPET level & EHS Safety Level   65
 A2.3 Full Time Equivalent (FTE) Resource Alignment   68
  A2.3.1 Process Hazard Management Coordinator and Hazard
         Assessment Facilitator Qualification   69
 A2.4 Expertise and Experience   69
 A2.5 EHS and PHM Alignment   69
 A2.6 Overview of Duties and Responsibilities   72
  A2.6.1 PHM Coordinator   72
  A2.6.2 Hazard Assessment Facilitator   72
 A2.7 Competency-Based Knowledge (Training) Road Map for Qualification   73
APPENDIX 3: HAZOP FACILITATOR   **75**
APPENDIX 4: SHOWING GAP CLOSURE PROGRESS   **79**
INDEX   **81**

LIST OF TABLES

TABLE 2.1.   Common Job Titles   14
TABLE 2.2.   Example of Skills and Knowledge of Safe Work Practices
             for Proficiency Levels   22
TABLE 3.1.   Customize the Tasks List in the RBPS Table   27
TABLE 3.2.   Modified Job Roles for RBPS Table   28
TABLE 3.3.   Modified Job Roles for RBPS Table for an Engineering
             Design Firm   28
TABLE 3.4.   Example of the RBPS Table Customized for Proficiency
             Requirements   30
TABLE 3.5.   Simplified Example Proficiency Requirements for HSE
             Manager   33
TABLE 3.6.   Example of Some of the Skills, Knowledge and/or
             Experience Requirements for HSE Manager   34
TABLE A1.1. Demonstration of Competency   64
TABLE A2.1. Hazard Levels Definitions   67
TABLE A2.2. Facility Level Assignment   68
TABLE A2.3. Facility PHM Requirements Based on Hazard Level   71
TABLE A2.4. Facility PHM Competency Requirements Based on
             Hazard Level   74
TABLE A3.1. Process Safety Element: Process Hazard Analysis (PHA)   75
TABLE A3.2. HAZOP Proficiency Level Example   76
TABLE A4.1. Gap Closure Progress   79

FILES ON THE WEB ACCOMPANYING THIS BOOK

Access Defining Process Safety Competency Requirements tools and documents using a web browser at:

http://www.aiche.org/ccps/pscomp

Password: Competency2014

ACRONYMS AND ABBREVIATIONS

AIChE	American Institute of Chemical Engineers
API	American Petroleum Institute
ASME	American Society of Mechanical Engineers
BEAC	Board of Environmental, Health, and Safety Auditor Certifications
BP	British Petroleum
BSI	British Standards Institution
CSB	US Chemical Safety and Hazard Investigation Board
CCPS	Center for Chemical Process Safety
CEO	Chief Executive Officer
CFR	Code of Federal Regulations
CPI	Chemical Process Industries
EFCE	European Federation of Chemical Engineering
EHS	Environmental, Health and Safety, also HSE or SHE
EPA	US Environmental Protection Agency
ERT	Emergency Response Team
FTE	Full Time Equivalent
HAZOP	Hazard and Operability
HazPET	Hazardous Process Evaluation Tool
HAZWOPER	Hazardous Waste Operations and Emergency Response
HSE	Health, Safety and Environmental
IChemE	Institution of Chemical Engineers
ISO	International Organization for Standardization
MOC	Management of Change
MOOC	Management of Organizational Change
NFPA	National Fire Protection Association
OCM	Organizational Change Management
OECD	Organization for Economic Co-operation and Development
OSHA	US Occupational Safety and Health Administration
PHA	Process Hazard Analysis
PHM	Process Hazards Management
PSCM	Process Safety Competency Matrix
PSM	Process Safety Management

RAGAGEP	Recognized and Generally Accepted Good Engineering Practices
RBPS	Risk Based Process Safety
RMP	Risk Management Plan
SAWS	State Administration of Work Safety
SHE	Safety, Health and Environmental
UK	United Kingdom
US	United States

GLOSSARY

Competency
A PSM program element associated with efforts to maintain, improve, and broaden knowledge and expertise

Job Role
A position within the typical company or organization that is responsible for a certain set of common tasks and activities (e.g., process safety).

Management of Change (MOC)
A management system to identify, review, and approve all modifications to equipment, procedures, raw materials, and processing conditions, other than replacement in kind, prior to implementation to help ensure that changes to processes are properly analyzed (for example, for potential adverse impacts), documented, and communicated to employees affected.

Organizational Change
Any change in position or responsibility within an organization or any change to an organizational policy or procedure that affects process safety.

Organizational Change Management (OCM)
A method of examining proposed changes in the structure or organization of a company (or unit thereof) to determine whether they may pose a threat to employee or contractor health and safety, the environment, or the surrounding populace.

Process Safety Management (PSM)
A management system that is focused on prevention of, preparedness for, mitigation of, response to, and restoration from catastrophic releases of chemicals or energy from a process associated with a facility.

Proficiency Level
The degree of expertise required for a specific subject area, is defined by how much experience, knowledge and skills someone needs to maintain in order to meet one of the following definitions:

Awareness   Is generally aware of the topic and associated terms. May not know the answer, but knows where to get more information.

Basic Knowledge Has general working knowledge of the topic. Has basic

	training required to perform general tasks related to their role.
Practitioner	Can execute specific tasks within the topic with minimal direction. Has the experience levels to complete assigned tasks. Is typically an engineer or manager.
Expert	Is a recognized expert with extensive knowledge and skills. Has specialized training or certification which may be required for certain tasks.
Leader	Is responsible for implementation of a PSM element or activity. Knowledgeable of the subject matter, may be a subject matter expert, but that is not required. Coordinates all resources and assigns tasks to ensure effective implementation of the topic.

ACKNOWLEDGMENTS

The American Institute of Chemical Engineers (AIChE) and the Center for Chemical Process Safety (CCPS) express their appreciation and gratitude to all members of the Process Safety Competency Project and their CCPS member companies for their generous support and technical contributions in the preparation of these Guidelines.

SUBCOMMITTEE MEMBERS:

Jeff Fox	Committee Chairman, Dow Corning Corporation
Mike Broadribb	Baker Engineering and Risk Consultants
Tom Burton	3M
Paul Delanoy	The Dow Chemical Company
S. Ganeshan	Toyo Engineering India
Kay Koslan	Dow Chemical
Mikelle Moore	Buckman
Eric Peterson	MMI Engineering
Sara Saxena	BP
Eddy Schedule	ABS Group Inc.
Jeff Stawicki	The Lubrizol Corporation
William Ward	Praxair Inc.

AIChE and CCPS also wish to express their appreciation to Melody Soderberg, Henry Ozog, Elena Prats and Molly Myers of ioMosaic Corporation for their contributions in preparing this book for publication and to Albert Ness of the CCPS for final editing. The collective industrial experience and know-how of the subcommittee members plus these individuals makes this book especially valuable to engineers who develop and manage process safety programs and management systems, including the identification of the competencies needed to create and maintain these systems.

Before publication, all CCPS books are subjected to a thorough peer review process. CCPS gratefully acknowledges the thoughtful comments and suggestions of the peer reviewers. Their work enhanced the accuracy and clarity of these guidelines.

Peer Reviewers:

Shirlyn Cummings	Bayer MaterialScience, LLC
Morgan Reed	MMI Engineering
John Remy	LyondellBasell
Emmanuelle Hagey	NOVA Chemicals Corporation
David Lewis	Occidental Chemical Corporation
Neil Maxson	Bayer MaterialScience, LLC
Louisa Nara	CCPS
Pamela Nelson	Cytec Industries
Robert Bussey	Merck & Company, Inc.
Walt Frank	Frank Risk Solutions
Cheryl Grounds	BP

Although the peer reviewers have provided many constructive comments and suggestions, they were not asked to endorse this book and were not shown the final manuscript before its release

PREFACE

The American Institute of Chemical Engineers (AIChE) has been closely involved with process safety and loss control issues in the chemical and allied industries for more than four decades. Through its strong ties with process designers, constructors, operators, safety professionals, and members of academia, AIChE has enhanced communications and fostered continuous improvement of the industry's high safety standards. AIChE publications and symposia have become information resources for those devoted to process safety and environmental protection.

AIChE created the Center for Chemical Process Safety (CCPS) in 1985 after the chemical disasters in Mexico City, Mexico, and Bhopal, India. The CCPS is chartered to develop and disseminate technical information for use in the prevention of major chemical accidents. The center is supported by more than 160 chemical process industries (CPI) sponsors who provide the necessary funding and professional guidance to its technical committees. The major product of CCPS activities has been a series of guidelines to assist those implementing various elements of a process safety and risk management system. This book is part of that series.

Process safety competency is a fundamental element of successful process safety programs and is closely related to the knowledge and training elements of the Risk Based Process Safety (RBPS) system. Facilities continue to be challenged with how to define process safety competency to improve organizational learning and process safety culture. The CCPS Technical Steering Committee initiated the creation of this guideline to assist facilities in meeting this challenge, and these guidelines describe how to use the matrix. This book contains approaches for developing, implementing, and continually improving a process safety competency matrix.

EXECUTIVE SUMMARY

Process safety practices and formal safety management systems have been in place in some companies for more than 100 years. Process safety management (PSM) is widely credited for reductions in major accident risk and for improved chemical Industry performance. Nevertheless, many organizations are still challenged with effectively implementing the management systems they have developed. In 2005 The Center for Chemical Process Safety introduced the concept of Risk Based Process Safety (RBPS) as a means to overcome what seemed like stagnation in the process safety journey. RBPS is described in more detail in section 1.4 of this book.

The purpose of this Guideline book is to help organizations design and implement the Process Safety Competency element of RBPS. This book provides ideas and methods on how to define job roles and proficiency levels within an organization, and then assign basic knowledge and skills to each job role/proficiency level combination through a Process Safety Competency Matrix (PSCM).

The PSCM enables better implementation of other elements of RBPS, especially Training and Performance Assurance, Management of Change and Organization Change Management (OCM), and Conduct of Operations.

The PCSM also enables an organization to achieve the goals of Vison 20/20, a CCPS initiative to help organizations achieve great process safety.

ORGANIZATION OF THIS BOOK

Chapter 1 of the book briefly describes how to use this book, provides a brief overview of some key elements RBPS, a discussion of what competency means and how competency fits into CCPS Vision 20/20. Chapter 2 describes how to classify roles in the organization, identify proficiency levels, and develop a list of skills and knowledge. Chapter 3 describes the PSCM and how to construct one. Starting examples of the PCSM are provided in the web site that is listed in the front of this book. Chapter 4 discusses how to assess existing competencies and also discusses competencies for process safety managers and Process Hazard Analysis (PHA) facilitators. Chapter 5 describes how to assess competencies, and Chapter 6 how to develop a plan to close gaps between the assessed competencies from Chapter 5 and the needed competencies from Chapter 3. Chapter 7 discusses strategies for maintaining competencies.

1

INTRODUCTION

1.1 WHY PROCESS SAFETY COMPETENCY?

The Process Industries successfully use hazardous materials and operate potentially hazardous processes safely every day. It is the very ability to do so that demonstrates the value of the process industries to society. The safe and successful operation of these processes is only accomplished by companies and individuals who have demonstrated the ability to carry out their tasks correctly and safely. This proper application of knowledge and skills demonstrates what we refer to as competency.

In times past, the apprentice often spent years training under the watchful eye of the master craftsman until the apprentice could show competency in the trade. Then, and only then would the apprentice be recognized with the title of craftsman. And so it is today, the process industries need to not only gain skills and knowledge to properly operate process equipment, but they must also be able to demonstrate the application of these skills and knowledge to fulfill the mission of the organization. This challenge continues throughout the lifecycle of the organization as individuals and roles change through time, so a structured approach to managing competency is needed.

This Guidance book was written to help organizations use a structured approach to define the necessary competencies needed to successfully fulfill the mission of the organization and meet the regulatory requirements in the jurisdiction in which it operates.

1.2 PURPOSE

The intent of this guideline book is to provide a framework defining the minimum recommended process safety knowledge and expertise (competence) necessary for a company or organization that handles hazardous materials to ensure that process safety is managed effectively. This book guideline does discuss how supporting personnel involved with engineering design and projects can modify the tools to best suit their use (see Section 2.1.11 *Additional Roles*). The main tool for conveying this information is through a Process Safety Competency Matrix (PSCM) or the Matrix. This Matrix is based on the principles outlined in the CCPS *Guidelines for Risk Based Process Safety* (RBPS) (CCPS 2007) Guidelines.

The Matrix forms the basis for ensuring that these competencies are maintained as changes are made in the organization (see CCPS *Guidelines for Managing Process Safety Risks During Organizational Change*) (CCPS 2013).

1.2.1 Address Applicable Regulations and Requirements

This book provides guidance on how to enhance existing Process Safety Management (PSM) programs that are designed to comply with regulations such as the OSHA PSM Standard (29 CFR 1910.119) in the United States, the Seveso Directive in Europe, the PSM regulation AQ/T 3034-2010 in China or the Singapore Standard SS 506 for Process Safety. It will also help companies that are implementing the Process Safety Competency and Management of Change (specifically Management of Organizational Change) elements of the RBPS Guidelines or the Process Safety Code under Responsible Care® (RCMS and RC14001 2013).

1.2.2 Identify Process Safety Competency Requirements

For facilities handling hazardous materials, the process-specific hazards can also present a need for process safety competency. This book provides guidance on how an organization can identify the necessary process safety competencies to manage the risks of handling hazardous materials. Levels of competency will likely align with the knowledge, skills and experience needed to manage hazards, based on the severity of the hazard and complexity of the process. Most process safety competencies will be associated with individuals located at an operating facility that handles hazardous chemicals. However, some process safety competencies may be associated with individuals at corporate or business unit offices, engineering or research centers, or outside consulting or engineering personnel.

1.2.3 Assess Individuals Against Recommended Competencies

After the recommended process safety competencies are determined, gaps between required training or experience levels can be identified. These gaps can then be filled by providing the necessary training, or by hiring new employees or contractors with the requisite competencies.

1.3 AUDIENCE

The target audience for this book is individuals who develop and manage process safety programs for either entire companies or an individual site. It includes corporate executives, business unit managers, corporate and site process safety managers, corporate and site process safety experts, operations managers, operators, mechanical integrity personnel, and other process safety personnel. It also includes process safety consultants and engineering companies. As evidenced by the extensive list, there is a need for awareness of process safety competencies throughout the organization, especially at the corporate level.

1.4 HOW TO USE THIS PROCESS

The PSCM provided in these guidelines provides a template for organizations to map their process safety job roles against those listed in the Matrix. It is not intended to prescribe how a process safety organization defines and develops individual competencies, but is intended to provide a template for organizing and managing process safety competencies and determining if there are any gaps.

The PSCM was developed using the four pillars and 20 elements of process safety as defined in the CCPS *Guidelines for Risk Based Process Safety* (CCPS 2007). For engineering design process safety competencies, additional guidance is included based on the CCPS *Guidelines for Engineering Design for Process Safety* (CCPS 2012). The five levels of proficiency are listed below:

- Awareness
- Basic Knowledge
- Practitioner
- Expert
- Leader

The Matrix is provided in a Microsoft ® Excel spreadsheet to be used for easy customization. There are multiple worksheet tabs of information that will be described throughout the chapters of the guidelines, therefore, it is strongly recommended to have the Matrix downloaded or printed while reading through this book. The Proficiency Levels tab further describes the five proficiency levels. The Skill and Knowledge tab provides guidance on the process safety competencies for each proficiency level for each **RBPS** element. The **RBPS** Table tab of the Matrix lists specific competency requirements for each RBPS element and the proficiency levels required for typical job roles in an organization. Once a company's job roles are mapped to the Matrix, gaps can be identified and an action plan developed to address these gaps.

1.5 RISK BASED PROCESS SAFETY ELEMENTS

Process safety practices and formal safety management systems have been in place in some companies for many years. Over the past 20 years, government mandates for formal process safety management systems in Europe, the U.S., and elsewhere have prompted widespread implementation of a management systems approach to process safety management. However, after an initial surge of activity, process safety management activities appear to have stagnated within many organizations. Incident investigations continue to identify inadequate management system performance as a key contributor to the incident. And audits reveal a history of repeat findings indicating chronic problems whose symptoms are fixed again and again without effectively addressing the technical and cultural root causes. This is one of the reasons the Center for Chemical Process Safety (CCPS) created the next generation process safety management framework – *Risk Based Process Safety (RBPS)*.

An RBPS management system consists of four main accident prevention pillars or foundational blocks, with twenty process safety elements under the blocks. The RBPS pillars and process safety elements are shown in Table 1.1

A thorough description of the RBPS pillars and elements can be found in the CCPS *Guidelines for Risk Based Process Safety* (CCPS 2007). Elements directly impacted by this book and the PSCM are covered in the following sections.

Table 1.1 RBPS Pillars and Process Safety Elements

Pillar	Process Safety Element
Commit to Process Safety	Process Safety Culture
	Compliance with Standards
	Process Safety Competency
	Workforce Involvement
	Stakeholder Outreach
Understand Hazards and Risk	Process Knowledge Management
	Hazard Identification and Risk Assessment
Manage Risk	Operating Procedures
	Safe Work Practices
	Asset Integrity and Reliability
	Contractor Management
	Training and Performance Assurance
	Management of Change
	Operational Readiness
	Conduct of Operations
	Emergency Management
Learn From Experience	Incident Investigation
	Measurement and Metrics
	Auditing
	Management Review and Continuous Improvement

In the next sections the process safety competency element, and a few selected elements that have an effect on or which process safety competency effects, are discussed in more detail.

1.5.1 Process Safety Competency

Competency is simply the ability of a person to do a job properly. Competency involves knowledge, skills and attitudes. The process safety competency element involves increasing the body of knowledge and, when applicable, pushing newly acquired knowledge out to appropriate parts of the organization, sometimes independently of any request. Most important, this element supports the application of the body of process knowledge to situations that help manage risk and improve plant performance. The main product of the competency element is an understanding and interpretation of knowledge that helps the organization make better decisions and increases the likelihood that individuals who are faced with an abnormal situation have the knowledge and skills to take the proper action.

Process safety competency enables proactive learning. One source of learnings comes from process safety incidents and near misses. Organizations are expected to apply learnings from incidents and near misses from one plant to the whole enterprise; in fact, this can be a regulatory requirement in some case. Another case where rapid dissemination of new knowledge can be needed is mergers and acquisitions. Roles in the new organization will change for many people. Positions may be combined, some may be eliminated. People in the new positions may have to acquire knowledge that was not pertinent to their old role or organization. An organization with a good process safety competency tool can ensure that learnings do get spread top the whole organization and remain part of the package of process safety knowledge.

The RBPS Guideline book does not describe how to achieve process safety competency. The PSCM tool described in this book provides a "how to" framework for an organization to develop process safety competency at all levels by identifying the knowledge and competency needed for all the job roles that are identified by an organization. Companies may have to modify the PSCM to fit their organization, however, with the framework established, this is a cut and paste exercise. Therefore the PSCM is an enabler of the goal of developing training programs and increasing process safety knowledge and getting the right knowledge to the right people.

1.5.2 Corporate Process Safety Culture

Process safety culture is the first element under the Commit to Process Safety pillar. Process safety culture is defined in the RBPS Guidelines as "the combination of group values and behaviors that determine the manner in which process safety is managed" (CCPS 2007). The process safety culture of an organization will have a major impact on how process safety issues are managed. A positive process safety culture can be achieved throughout an organization if direction comes from top management. This is typically communicated to the organization through a vision statement, a set of process safety goals and objectives, and supported by appropriate behavioral-based, reinforcing safety programs. Management commits to provide the resources to meet the goals and objectives and ensure that unsafe behavior is not tolerated

Competency, as described in section 1.2.4, has an attitude component. The competency tool described in this book reinforces a good process safety culture by demonstrating the organization cares enough about process safety to take the time to establish a PSCM and follow through with the training and other implications of the matrix. The strong direction from top management supports leaders' at all organizational levels drive a positive process safety culture. Safety culture leaders are not just defined by their position, but rather are personnel who have the inherent characteristic to uphold safe actions, and influence others to do the same, leading the positive progression of safety culture.

Once a positive process safety culture is established, all employees will understand their responsibilities and deeply-held values will be reflected in the actions of individuals. New employees will quickly understand these values and adopt them as their own. Individual and group process safety achievements should be recognized and rewarded. Individuals may have process safety goals as part of their individual goals and objectives which are reviewed during performance reviews.

An organization's process safety values and expectations may be apparent to many employees, but to be effective, these values and expectations must be communicated clearly to everyone. This may include a formal training program which includes the company process safety vision, goals and expectations. Also the training should emphasize the need for process safety competency at all levels of the organization and the need to maintain these competencies through all types of organizational changes, ranging from retirement of process safety experts to elimination of entire layers of management or supervision. A key component of a positive safety culture is to have all required process safety competencies identified, implemented and managed at all times, especially when there are organizational changes.

1.5.3 Process Safety Knowledge Management

The *process safety knowledge* element involves work activities associated with compiling, cataloging, and making available a specific set of data that is normally recorded in paper or electronic format. However, *knowledge* implies understanding, not simply compiling data. In that respect, the *competency* element complements the *knowledge* element in that it helps ensure that users can properly interpret and understand the information that is collected as part of this element.

The *knowledge* and *competency* elements are closely linked in other ways as well. The *competency* element involves work activities that promote personal and organizational learning and help ensure that the organization retains critical information in its collective memory. Technology manuals and other written documents produced as part of the *knowledge* element are often stored and distributed via the *competency* element's management system. A technology manual compiled as part of the *competency* element often includes historical information; it is also less structured and is much more likely to include sections that are "under development" based on ongoing projects within the company's research and technology functions.

1.5.3 Organizational Change Management (OCM)

Management of Change (MOC) is an element under the Manage Risk pillar. The importance of MOC may not be new to most organizations today, but the importance of OCM, also known as Management of Organizational Change (MOOC), is less well understood. Organizational change covers changes in working conditions, personnel changes, task allocation changes, organizational structure changes and policy changes that can affect process safety. The CCPS *Guidelines for Managing Process Safety During Organizational Change* (CCPS 2013) covers this subject in more detail.

MOOC and process safety competency are closely linked. The cornerstone of an MOOC program is an established process safety competency element. To create an MOOC management process, the roles and responsibilities of individuals with respect to process safety must be clearly defined and a process in place to obtain the appropriate level of process safety knowledge in every individual. As discussed in section 1.5, The PSCM is the tool that enables the process safety competency element.

1.6 CCPS VISION 20/20

CCPS Vision 20/20 looks to the future to describe how great process safety is delivered. There are five tenets for industry in Vision 20/20:

- Committed Culture
- Vibrant Management Systems
- Disciplined Adherence to Process Standards
- Intentional Competency Development
- Enhanced Application of Sharing and Lessons Learned

The relevance of the PSCM to the *Intentional Competency Development* tenet is straightforward. Use of the PSCM ensures all employees are capable of meeting the technical and behavioral requirements of their jobs. The PSCM enables the other tenets as well.

In a *Committed Culture* all employees, from executives to employees on the plant floor demonstrate a commitment to process safety. Establishing a PSCM demonstrates that commitment.

Vibrant Management Systems call for all employees to understand their roles in managing process safety. The PSCM enables this by defining those roles.

Disciplined Adherence to Standards is enabled by defining where the responsibility for understanding, maintaining and disseminating standards appropriate to the organization.

Enhanced Application and Sharing of Lessons Learned can be enabled by first defining who is responsible for spreading lessons learned and by showing what training programs need to be updated as new knowledge becomes available.

1.7 REFERENCES

CCPS 2007. Center for Chemical Process Safety (CCPS), Guidelines for Risk Based Process Safety, New York, 2007.

CCPS 2013. Center for Chemical Process Safety (CCPS), Guidelines for Managing Process Safety Risks During Organizational Change, New York, 2013.

CCPS 2012. Center for Chemical Process Safety (CCPS), *Guidelines for Engineering Design for Process Safety*, New York, 2012.

RCMS. Responsible Care Management System® (RCMS), American Chemistry Council, Washington, DC.
<http://responsiblecare.americanchemistry.com/Responsible-Care-Program-Elements/Management-System-and-Certifcation/default.aspx>

RC14001 2013, *Technical Specification* (2013 edition), American Chemistry Council, Washington, DC.

CCPS 2013, Center for Chemical Process Safety (CCPS), *Guidelines for Managing Process Safety During Organizational Change*, New York, 201.

2

IDENTIFY PROCESS SAFETY ROLES & COMPETENCY NEEDS

Every organization has a unique hierarchy and structure which best suits its size, culture and purpose. Sometimes different facilities within the same organization are structured differently from one another. In order to understand and utilize the Process Safety Competency Matrix (PSCM), it is necessary to adapt the matrix to fit the specific organization and/or facility. This chapter defines the framework of the PSCM. It is necessary to have an understanding of the components of the Matrix template prior to customizing it for use at a specific facility.

2.1 LIST OF GENERIC JOB ROLES

The PSCM is organized using job roles that are common to most organizations. However, not every organization uses the same terminology for job roles and functions. Additionally, different sizes of organizations may either have consolidated some of these job roles or they may have many more job roles. Therefore, an explanation of each of the generic job roles is included to enable a mapping of these common roles to the actual titles used within a specific organization. Table 2.1 *Common Job Titles*, includes a cross reference for some common alternative job titles. Following Table 2.1, explanations and typical responsibilities are provided.

TABLE 2.1.   Common Job Titles

Title Used in RBPS Table	Possible Alternative Titles
Facility Manager	Plant Manager/ Leader
	Facility Manager
	Site Superintendent
	Site Manager/Leader
	Site Director
HSE Manager	Safety Manager
	EHS Manager/ Leader
	SHE Manager
Operations Manager	Site/Facility Superintendent
	Production Manager
Inspection, Testing, and Maintenance Manager	Risk & Reliability Manager
	Asset Integrity Manager
	Maintenance Manager
	Inspection Manager
Technician	Mechanic or Operator
Supervisor	Foreman
	First Line Supervisor
	Shift Leader

2.1.1 Facility Manager

The Facility Manager is the top leadership position at a site location of a company. If this is a single manufacturing site within a larger organization which runs multiple manufacturing facilities, then this would be the leader for a particular site. If the organization is small and only has a single manufacturing facility, this could be the company president or CEO.

2.1.2 HSE Manager

The HSE Manager is the person at the facility who has been assigned responsibility for process safety. This role is often combined with managing all aspects of safety, including personnel and process safety. Sometimes this role is also consolidated with health (i.e., industrial hygiene), environmental compliance, and/or security.

In some companies, the Process Safety Manager is a separate job role from the HSE Manager and reports through the engineering, technical, or operations department. The PSCM can be modified to accommodate the two roles by adding an additional column and separating the competencies according to the responsibilities of each manager. Table 2.1 depicted below, provides typical titles used in the PSCM and possible alternative titles.

2.1.3 Operations Manager

The Operations Manager is typically the highest ranking person at the facility who is in charge of production or manufacturing. This person commonly reports to the Facility Manager. If the facility has multiple production units, this person would have responsibility for all of the different units. If the facility is small and only has a single production unit, this role may be the same as the Facility Manager.

2.1.4 Inspection, Testing and Maintenance Manager

The Inspection, Testing and Maintenance Manager is the person who is in charge of the inspection, testing, and maintenance functions for the facility. In some organizations, this may be a single department and in other organizations the inspection and testing department is completely separate from the maintenance department. This person(s) commonly is responsible for ensuring that policies and procedures are in place for inspection, testing, and repairs on all process equipment. Sometimes there may be one person in charge of the mechanical items and another person in charge of the instrumentation and controls items.

The PSCM can be modified to accommodate how these responsibilities are managed by adding additional columns for separate job roles and aligning the competencies according to the responsibilities of job role.

2.1.5 Supervisor

A Supervisor would typically be the first level of management above a group of operators, technicians, or employees with no personnel management responsibilities. In some organizations, this could also be an operator who has a broader management role. The person in this role would typically supervise a group of operators or technicians who run the day-to-day operations at the facility. This person is commonly in charge of scheduling, training, and coaching the personnel who report to him/her.

2.1.6 Operator

An Operator is the individual who runs and monitors the chemical process at the facility. This would include personnel who interface with the process control system as well as those who handle the field-based tasks required to safely produce a product. It could also include individuals who handle sampling of process streams, or otherwise interact directly with the process.

2.1.7 Technician

A Technician is the individual who handles inspections, tests, or maintenance of equipment. This would include personnel who handle all types of equipment including vessels, rotating equipment, instrumentation and controls. It could also include personnel who handle the day-to-day maintenance on the basic process control system.

2.1.8 Engineer

This role may provide expertise in specific aspects of engineering and process safety, and may be responsible for implementation of process safety. It may sometimes manage process safety documentation for the area. It could also include personnel involved with process development to process improvement and troubleshooting.

2.1.9 Project Manager

This role manages the design, installation, and start-up of new process equipment or smaller changes managed through the management of change process. This role could be held by someone assigned to the local facility, or someone from a corporate support role. This role would also apply to someone at an engineering design or construction company that may be hired to design and install new process equipment and/or facilities.

2.1.10 Project Engineer

Like the Project Manager, this person would be involved in the execution of the process or equipment project, and would organize the resource utilization and resource mobilization. This role could be held internal to the organization such as someone assigned to the local facility or someone from a corporate support role, or externally, such as someone in contracted engineering design firm or construction company. A Project Engineer who is highly involved in the design aspects may be considered the design engineer. A Project Engineer must have a thorough understanding of the codes and standards applicable to their work to ensure that the new equipment is in compliance and can be operated safely.

2.1.11 Additional Roles

Although this skills matrix is primarily organized with a manufacturing facility in mind, it can be easily adapted for other types of organizations, such as engineering design firms. If an organization has a very different structure than the one used for the Matrix template, it may be necessary to insert additional columns for the new roles which are applicable to the organization and remove the columns for roles which are not present in the organization. An example of this for an engineering design firm would be to add new columns for each of the engineering disciplines (i.e., expand the Projects section), and remove the columns for Operations and Inspection, Testing and Maintenance.

2.2 LIST OF PROFICIENCY LEVELS

The PSCM template was developed with five proficiency levels. This provides a clear separation of proficiencies across an organization. However, some organizations, particularly smaller ones, may opt to consolidate some levels to better match their organizational structure. If fewer proficiency levels are desired, it is suggested to combine the "Basic Knowledge" and "Practitioner" levels together.

For larger organizations it may be desirable to expand the proficiency levels.   In this case, there might be a "Master" proficiency level which is higher than the "Expert", who is basically the expert within the company that all others look to for setting direction and guidance. Higher proficiency levels will typically include all proficiencies for lower levels.

2.2.1 Awareness Proficiency Level (Level 1)

The "Awareness" proficiency level is the most common. It requires a general overview of process safety and the elements from which it is comprised.

All personnel at the facility, including direct employees and contractors, who could affect process safety should, as a minimum, have an awareness of process safety elements that are relevant to their job. Without a basic understanding of process safety, facility personnel may not understand how their actions may inadvertently put people or the facility at risk. With this entry level of proficiency, personnel would be expected to understand the hazards of the process and who they could approach for additional information.

Some of the process safety elements aren't relevant to certain job roles. For example, a process operator would NOT be expected to be aware of the principles of inherently safer design (ISD) or other process safety elements related to engineering design. However, other job roles involved with designing modifications and projects, such as a project engineer, should, as a minimum, have an awareness of ISD and other engineering elements, and may possess a higher proficiency level. The "Awareness" level sets an expectation that all personnel know what process safety elements are relevant to their job. Because everyone is expected to have at least an awareness level of proficiency, for elements that are relevant to their job, "Awareness" does not appear in the RBPS table of the Matrix, in order to avoid clutter. If desired, this proficiency level may be inserted where relevant for all roles in the RBPS table.

2.2.2 Basic Knowledge Proficiency Level (Level 2)

The next level of proficiency and the first one designated in the RBPS table is the "Basic Knowledge" level. Personnel meeting this proficiency level would have a general working knowledge of the topic and basic training to carry out general process safety tasks related to their role. This proficiency level is generally required for anyone who will be expected to participate, in some way, with implementing a process safety element, such as operators and technicians. This would typically include personnel who may be team members, such as a Process Hazard Analysis (PHA) or hazard assessment team, incident investigation team, pre-startup safety review team or other process safety committee. This proficiency level would also include personnel who need to be able to carry out actions assigned to them for a process safety element such as using operating procedures or safe work permits.

2.2.3 Practitioner Proficiency Level (Level 3)

This proficiency level builds upon the "Basic Knowledge" level. The "Practitioner" level of proficiency is the one typically assigned to engineers and HSE managers. These would be the personnel who would typically handle most of the tasks associated with implementing each process safety element. These people need to have a thorough working knowledge of the assigned process safety element and enough experience to carry out the majority of the work needed for compliance.

2.2.4 Expert Proficiency Level (Level 4)

An expert can either be seen as a final decision maker for any decisions regarding an area of process safety competency, or a subject matter expert who is relied upon to answer questions about a process safety element.

This proficiency level builds upon the "Practitioner" level. This proficiency level is for sub-elements that need a high level of experience or expertise. In some situations, personnel may need to have specific certifications or training.

These certifications may be obtained from various technical organizations such as ISO, ASME, IChemE and API. As of this writing, CCPS is also considering a certification program.

Not all process safety elements require personnel to have this level of proficiency. Not every company needs an expert. An organization may decide whether an expert is needed in an area of process safety competency. There are experts available throughout the industry in all elements of process safety that could be called upon to provide additional guidance, if needed.

2.2.5 Leader Proficiency Level (Level 5)

The "Leader" proficiency level is typically held by the person assigned to the overall responsibility for implementing a particular process safety element. This proficiency level does not fit within the progression of the other four levels of proficiency. Because the person assigned to this role is responsible for implementation, this proficiency level would include knowledge, skills, and experience related to management of personnel and documentation which may be required for compliance.

2.3 LIST OF PROCESS SAFETY KNOWLEDGE/SKILLS

The Skills & Knowledge tab in the Matrix provides guidance on typical competency levels required for each element and each proficiency level. These are measurable criteria, illustrated in Table 2.2 *Example of Skills and Knowledge of Safe Work Practices for Proficiency Levels*, by which personnel in assigned roles with a required competency level can be compared. These criteria consist of knowledge, skills and experience, or a combination of these for each process safety element and proficiency level.

TABLE 2.2. Example of Skills and Knowledge of Safe Work Practices for Proficiency Levels

Pillar	Element	Basic	Practitioner (Basic Level Plus the Following)	Expert (Practitioner Level Plus the Following)	Leader
III. Manage Risk	Safe Work Practices	Basic knowledge of company policies with regard to safe work practices Knowledge on how to use safe work practices that may be needed	Knowledge of, or experience with, issuing safe work permits that may be needed Knowledge and experience necessary to train employees on how to properly implement safe work permits	Knowledge of regulatory requirements for implementing safe work practices Knowledge and experience to audit compliance with safe work practices	Responsible for implementation of a PSM element or activity Knowledgeable in regulatory requirements and company procedures for managing safe work practices

3

PROCESS SAFETY COMPETENCY MATRIX

3.1 WHAT IS THE MATRIX?

The Process Safety Competency Matrix is a tool which can be used to identify and document the process safety competencies that are essential within an organization. It provides a framework to organize the requirements and assess needs for improvement. The Matrix should NOT be viewed as a definitive listing of requirements for all organizations. The Matrix is intended to be customized to fit each individual organization's needs.

The Matrix consists of several Microsoft ® Excel worksheets which work together. The Matrix is based on the *Guidelines for Risk Based Process Safety* (CCPS 2007) from CCPS and the four pillars and the 20 elements of process safety identified in that book. Both the RBPS and the Skills and Knowledge worksheets are organized around the pillars and elements. At the bottom of each of these worksheets there is additional guidance for Engineering Design which is based on the *Guidelines for Engineering Design for Process Safety* (CCPS 2013) from CCPS. This may be used as a guide for in-house project design groups or for separate engineering design firms.

The RBPS worksheet provides details on many of the key work activities required by each process safety element. For each element, a leader is identified in this worksheet. This leader is the person responsible for ensuring adequate execution of all activities associated with that element. The RBPS table also includes a listing of typical roles within an organization that may be called upon to conduct actions associated with a process safety element. For each of the process safety elements, there is an assignment of proficiency level required for all organizational roles that are needed to support that element. Note that some elements may not be applicable to a certain job role.

The Proficiency Level worksheet describes the five different proficiency levels which may be used to assign skill and knowledge requirements to properly implement each element. These proficiency levels have been described in detail in Chapter 2. The initial "Awareness" proficiency level is the base level. All personnel at the facility, including direct employees and contractors who could affect process safety should, as a minimum, have an awareness of process safety elements that are relevant to their job. Therefore, this level is not specifically called out in the Matrix RBPS table. The next three sequential proficiency levels are used for the variety of roles within the organization depending on the needs and structure of the organization. The "Leader" proficiency level is needed for the person who is assigned overall responsibility for an element in the RBPS worksheet.

The Skills and Knowledge worksheet describes the typical skills, knowledge, experience and/or certifications which may be needed for each of the different proficiency levels, organized by element. As mentioned earlier, the Awareness level of proficiency is not included in this table. Each of the three main levels ("Basic Knowledge", "Practitioner" and "Expert") build upon each other with each higher level needing the skills and knowledge of the lower levels plus the new requirements that are listed for that level.

The "Leader" proficiency level is not directly related to the other sequential levels and builds upon the "Basic Knowledge" level. In some instances, the leader assigned for a process safety element may require a proficiency level higher than "Basic Knowledge". This higher level of proficiency can be noted in the skills and knowledge column heading "Leader" ("Basic Knowledge" level plus the following).

3.2 HOW TO CUSTOMIZE THE MATRIX

3.2.1 The Proficiency Levels Worksheet

The starting point for utilizing the PSCM is the Proficiency Levels worksheet. The descriptions of these proficiency levels can be reviewed by the organization to determine if the Matrix needs to be adapted. A very small or flat organization may opt to combine the "Basic Knowledge" and "Practitioner" levels of proficiency.

A very large organization with multiple facilities and corporate support functions may want to add an additional "Master" proficiency level above the "Expert" level to recognize the key personnel who provide guidance to the experts for that element within the organization.

3.2.2 The Risk Based Process Safety Worksheet

The next step is to customize the RBPS worksheet. The outline buttons on the far left side of the worksheet can be used to expand each element to reveal the tasks for that Element in column C of the worksheet. These key tasks are based on the possible work activities in the *Guidelines for Risk Based Process Safety* book. An organization may need to customize these required tasks to reflect how they manage each process safety element, taking into account that sections of each element can be managed by multiple job roles. Refer to Table 3.1 for an example of how this list of tasks may be modified or expanded.

Once the tasks for each element have been modified as needed, a leader should be assigned for each element, and all the competencies within that element. The draft Matrix has tentative "Leader" assignments based on common organizational structures. However, each organization may choose to assign these responsibilities differently. The important thing is to ensure that each process safety element has a clear leader. Depending on preference, the "Leader" assignment can be identified by a role or job title or it can be a specific person. Refer to Section 2.2.5 *Leader Proficiency Level (Level 5)*, for additional details on what is expected for an element Leader.

TABLE 3.1.   Customize the Tasks List in the RBPS Table

Pillar	Element	Individual is competent to implement the following tasks
III. Manage Risk	Operating Procedures	Develop and implement written policy for creating and managing operating procedures (Procedure HSE 10).
III. Manage Risk	Operating Procedures	Develop operating procedures based on collaboration with operators.
III. Manage Risk	Operating Procedures	Coordinate operating procedures with operator training (Ensure that new operator training manuals are continually updated to match any revised operating procedures).
III. Manage Risk	Operating Procedures	Examine operator knowledge of procedural tasks to demonstrate comprehension of operating procedures.
III. Manage Risk	Operating Procedures	Periodically review operating procedures to ensure compliance with regulatory issues (reviews held annually to coincide with the operating procedure certification).
III. Manage Risk	Operating Procedures	Periodically review operating procedures to ensure that they are complete and accurately reflect the actual practice.
III. Manage Risk	Operating Procedures	Audit operator tasks to ensure that current practice reflects operating procedures.
III. Manage Risk	Operating Procedures	Establish a system to manage safe operating limits (use the Operating Parameter Tables for each process).

The next step is to review the roles listed at the top of the RBPS worksheet in rows 3 and 4. Each organization is unique and may utilize different titles for common roles. Table 2.1 *Common Job Titles*, has some common alternative job titles for many of these roles. These titles should be modified to reflect the organizational chart. Additionally, there may be additional or fewer roles than what are listed in this worksheet. These roles should be consolidated or expanded, as needed. If desired, the "Others" role in column Q may be expanded to more explicitly designate proficiency requirements. Table 3.2 illustrates an example of how a portion of the table has been modified for a different organization.

For non-production facilities, the roles across the top of the table may need to be significantly changed. For example, an engineering design firm may not have any Operations roles and their Inspection, Testing and Maintenance roles may be more focused on initial equipment inspections. In these instances, remove the roles and categories of roles which do not exist within the organization and expand the listing to encompass the roles which are critical to the organization.

Table 3.3 shows an example of how the roles section of the RBPS tab might be modified for an engineering design firm.

Once the roles have been adjusted, the last step is to assign proficiency levels for each element to the various roles within the organization. A review of the activities in column C can be helpful in assigning competencies. The assignments included in the Matrix template are based on typical organizations, but may not match an individual organization.

TABLE 3.2.   Modified Job Roles for RBPS Table

Production				Maintenance & Reliability			
Opera-tor	Shift Foreman	Process Engineer	Operations Manager	Mechan-ic	Supervi-sor	E&I or Mechanical Engineer	Mainte-nance Manager

TABLE 3.3.   Modified Job Roles for RBPS Table for an Engineering Design Firm

Engineering Disciplines (Engineers)					Management		
Process	Piping & Mechani-cal	Electrical	Instrument & Controls	Civil & Structural	Office Manager	Project Manager	Discipline Lead

All of the elements will typically need one or more roles assigned to a "Basic Knowledge" proficiency level and one or more roles assigned to a "Practitioner" proficiency level. Not all elements require an "Expert" proficiency level within the organization. The "Expert" proficiency level is typically needed when there are specific requirements for special certifications, experience, or training for a particular element. The "Expert" level may also be needed based on the complexity of the process and the severity of the process hazards. A company may choose to designate an "Expert" level person for all elements. Keep in mind that there are resources available for all of the process safety elements at the Expert Level, if they should be needed, on a contract basis. This is sometimes necessary to establish a strong foundation for a particular element or to significantly improve the performance in a particular element if a deficiency is noted.

Note that, although a role is assigned a proficiency level, it doesn't mean that everyone assigned to that role has to have that same proficiency level. In some instances, the organization may only need one or a few people within that role category to possess the proficiency level which is listed in the RBPS worksheet. In those cases, it may be helpful to add some notes to the customized table to reflect how many people require which proficiency level. That can add significant flexibility when making organizational changes if personnel assigned to a particular role continually develop experience and expertise to fill these process safety requirements. An example of clarifying the required proficiency numbers is shown in the excerpt of the RBPS tab in Table 3.4.

TABLE 3.4.   Example of the RBPS Table Customized for Proficiency Requirements

Element	Operations		HSE	
	Operator	**Engineer**	**Engineer**	**Manager**
Hazard Identification & Risk Analysis	Level 2 (Lead Operator and most senior person on each shift only)	Level 3 (Lead process engineer for each department only)	Level 4 (One Expert required)	Level 4 (If there are at least two other Expert Level personnel in the facility, then a
	Level 1 (All other operators)	Level 2 (all other process engineers in a department)	Level 3 (All other HSE Engineers)	Level 3 Proficiency is sufficient)

3.2.3   The Skills and Knowledge Worksheet

For each process safety element, the Skills and Knowledge worksheet outlines measurable benchmarks necessary to achieve each of the proficiency levels. In many instances the training required for some of the more basic proficiency levels can be done in-house through either formal training sessions or more informal one-on-one training depending on which option can meet the needs in the most efficient and timely manner. For the "Expert" proficiency level, specific outside training courses, work experiences or certifications are typically needed. To make it easier to measure, it may be helpful to edit the Skills and Knowledge worksheet to include specific training course titles and/or numbers where they are known.

In many cases the proficiency level has specific experience requirements. It may be beneficial to quantify this experience level more specifically to match the organization's expectations.

To reach the "Practitioner" or "Leader" proficiency level, the person must meet both the requirements listed for the "Basic Knowledge" proficiency level and the requirements listed for either the "Practitioner" or the "Leader" proficiency level. To reach the "Expert" proficiency level, the person must meet all of the requirements listed for the "Basic Knowledge", "Practitioner", and "Expert" levels.

If your organization opted to consolidate the "Basic Knowledge" and "Practitioner" proficiency levels, then the requirements for these two levels in the Skills and Knowledge worksheet should also be consolidated.

3.3 Uses of the Matrix

Once the Matrix is configured to suit the organization, the Matrix can be used to manage process safety competency and organizational change management.

A typical use will be to establish a training Matrix for all employees with process safety responsibilities. Another valuable use would be to assist in evaluating the impact of an organizational change. Both of these topics will be examined in the following sections. The Matrix may also be helpful in establishing requirements for job postings.

3.3.1 Establishing a Training Matrix

The first step in establishing a training matrix is to select the position or role to be evaluated. Locate the individual role in the RBPS table among the roles listed at the top. Highlight the proficiency level requirements for that column in the table. Review the listing of leader assignments within the RBPS table to flag any leader proficiency requirements that may be needed for that role or position. This summarizes the proficiency level requirements for that position or role. Refer to Table 3.5 for an example of the proficiency level requirements for a HSE Manager using the competency Matrix template.

In this table, some of the elements have no proficiency level required (shown as "none"). This example is taken from the PSCM where any blank boxes on the Matrix may represent the "Awareness" proficiency level (level "1"). The degree of awareness that each job role is expected to have for process safety elements is not specified on the template Matrix. This can be decided during the customization of the Matrix.

A level "2" represents the "Basic Knowledge" proficiency level, or a general knowledge of the element and how it applies to an HSE Manager. Level "3" is "Practitioner" and "4" is "Expert".

Some of these elements can be sub-divided. For example, Asset Integrity and Reliability has sub-elements, such as Inspection, Testing, Preventive Maintenance, Breakdown Maintenance, Reliability, Materials of Construction, Corrosion Management, QA/QC, Planning, etc. These sub-elements can also be divided into sub-components to differentiate between disciplines, such as rotating machinery, piping, electrical, control, structural, pressure vessels, etc. An HSE Manager is unlikely to understand all of these facets at the basic level; however they can understand how the general tasks are related to their position. Since this Matrix is customizable, it can be sub-divided to include these facets.

TABLE 3.5.   Simplified Example Proficiency Requirements for HSE Manager

Pillar	Element	Leader (Y/N)	Proficiency Level Required
I. Commit to Process Safety	Process Safety Culture	No	3
I. Commit to Process Safety	Compliance with Standards	Yes	3
I. Commit to Process Safety	Process Safety Competency	No	None
I. Commit to Process Safety	Workforce Involvement	Yes	3
I. Commit to Process Safety	Stakeholder Outreach	Yes	3
II. Understand Hazards and Risks	Process Knowledge Management	No	2
II. Understand Hazards and Risks	Hazard Identification and Risk Analysis	Yes	4
III. Manage Risk	Operating Procedures	No	None
III. Manage Risk	Safe Work Practices	Yes	4
III. Manage Risk	Asset Integrity and Reliability	No	2
III. Manage Risk	Contractor Management	Yes	3
III. Manage Risk	Training and Performance Assurance	No	3
III. Manage Risk	Management of Change	Yes	3
III. Manage Risk	Operational Readiness	No	3
III. Manage Risk	Conduct of Operations	No	2
III. Manage Risk	Emergency Management	Yes	4
IV. Learn From Experience	Incident Investigation	Yes	4
IV. Learn From Experience	Measurement and Metrics	Yes	3
IV. Learn From Experience	Auditing	Yes	3
IV. Learn From Experience	Management Review and Continuous Improvement	Yes	3
N/A	Engineering Design	No	None

Next, turn to the "Skills & Knowledge" worksheet and select the proficiency levels which match the requirements just identified. If a proficiency level is more than "Basic Knowledge", be sure to flag the required proficiency level for each element PLUS those lower proficiency levels which are also required (refer to section 3.2.3 for additional details). In some cases the skills from a higher level build upon those from a lower level so it is only necessary to list the highest requirement for a particular skill. Also, in some cases the skills required for the Leader proficiency level overlaps with the requirements for another proficiency level so these duplications would need to be removed. The table depicted below summarizes all of the required skills, knowledge, and/or experience needed to fulfill this role or position. Refer to Table 3.6 for an example of some of the required skills, knowledge and/or experience requirements for a HSE Manager using the competency Matrix template.

If the RBPS table has identified a potential for different proficiency levels for a particular role and element combination, be sure to highlight that fact when noting the proficiency level requirements. If there are multiple people filling the same role, then it may be helpful to note the higher proficiency level skills and knowledge as "optional", unless it is clear which personnel will have the higher proficiency level. This exercise may also be a good way to identify potential growth or development areas where personnel in this role may want to focus.

Once the listing of skills, knowledge, and/or experience is complete, begin comparing this against the skills, knowledge and/or experience that the personnel assigned to this role or position may have. Where gaps are identified, develop an action plan to close the gaps.

TABLE 3.6. Example of Some of the Skills, Knowledge and/or Experience Requirements for HSE Manager

Element	Proficiency Required	Skills, Training and/ or Experience Required
Emergency Management	4 & L	- Knowledge or experience in updating Emergency Response Plans based on learnings from drills - Experience in addressing findings from emergency response drills - Training and certification as an incident Commander - Training and certification as an ERT Member - Training and certification based on regulatory requirements (e.g.,HAZWOPER – US) - Ability to and/or certification to inspect, maintain and repair emergency response equipment - Knowledge of staffing requirements for emergency management - Ability to and experience with planning, executing, and critiquing emergency response drills
Incident Investigation	4 & L	- Knowledge of how to report incidents and near misses - Knowledge of appropriate role for incident investigations and follow-up - Knowledge of and/or certification in appropriate methodologies used to investigate incidents - Ability to and experience with how to properly interview personnel involved in or witnesses to incidents - Knowledge of documentation requirements for incident investigation - Knowledge of how to establish an incident investigation team - Knowledge of staffing requirements for an incident investigation team

Element	Proficiency Required	Skills, Training and/ or Experience Required
Measurement and Metrics	3 & L	- Knowledge of and experience on how to establish and measure leading and lagging process safety indicators - Knowledge of how to document process safety metrics
Auditing	3 & L	- Familiarity with PSM audits and knowledge of how to appropriately answer auditors questions - Experience with the requirements and information needed for auditing process safety programs - Knowledge of how to manage recommendations arising from audits - Knowledge of how to coordinate an audit
Management Review and Continuous Improvement	3 & L	- Knowledge of how to address recommendations from management reviews - Ability to and experience with interpreting trends of process safety performance - Knowledge of documentation requirements for management review and continuous improvement
Engineering Design	None	None

Maintain the listing of skills/knowledge and/or experience requirements and current status as a training matrix for that person. As the person develops, be sure to update the status on this training matrix.

If the person has additional skills, knowledge and/or experience beyond that required for the currently assigned proficiency level, note that in the training matrix. This may be helpful in the event of an organizational change or if the person is working towards a promotion or alternate position which may have different requirements.

3.3.2 Organizational Changes

An organizational change can take many different forms: working conditions, personnel changes, task allocation changes, organizational structure changes and policy changes. A risk assessment can be done to ensure that the competency levels within the organization are maintained (see *CCPS Guidelines for Managing Process Safety Risks During Organizational Change* (CCPS 2013a)). This competency matrix framework provides an excellent starting point for that risk assessment.

If the change just involved personnel changing roles, but the roles themselves are not changed, use the methodology outlined in Section 3.3.1 *Establishing a Training Matrix* to identify the required proficiency levels for any roles involved in the change. Then compare the skills, knowledge, and experience levels of the personnel who will be assigned to those roles against the requirements identified using the Matrix. Develop action plans to close any identified gaps, just as discussed earlier in this chapter.

This would be applicable, for example, when a person gets replaced and he/she had skills that were not required for the position, but others relied upon those skills. The new employee meets the process safety competency of the position, however a gap assessment would identify these additional skills and a gap closure plan would ensure that organizational change does not result in a loss of a skill-set. More about gap assessments and gap closure plans is discussed in Chapters 5 and 6.

If the change is more extensive and involves changes to the organization structure or the roles themselves, the Matrix can again be used to conduct a risk assessment for the change. In this case, carefully examine the RBPS table to ensure that all competencies are accounted for correctly with the new organization. To do this, save a copy of the current RBPS table for reference showing how the competencies are currently assigned within the organization. Next modify a copy of the Matrix to reflect the new set of roles within the organization and go through assigning proficiency levels for each element among the roles present in the revised organization. If roles are eliminated, ensure that there are sufficient assignments within the new organization to cover the required competencies adequately. This may mean that some roles within the organization may need to take on additional responsibilities. Whenever there are changes in proficiency level assigned to an existing role, if the new proficiency requirement is higher than the previous designation, then this may represent a gap in skills, knowledge and/or experience that must be addressed as part of the management of change process. For any new roles identified, develop the list of required proficiency levels and skills, knowledge and/or experience levels required for that new role, just as described in Section 3.3.1 *Establishing a Training Matrix.* This will serve as the basis for evaluating how well personnel proposed for this new role measure up to the requirements for the new role.

3.4   REFERENCES

CCPS 2007. Center for Chemical Process Safety, *Guidelines for Risk Based Process Safety*, New York, 2007.

CCPS 2013. Center for Chemical Process Safety, *Guidelines for Engineering Design for Process Safety*, New York, 2013.

CCPS 2013a. Center for Chemical Process Safety, Guidelines for Managing Process Safety Risks During Organizational Change, New York, 2013.

4

INDIVIDUAL AND ORGANIZATIONAL PROCESS SAFETY COMPETENCIES

4.1 DEVELOP ORGANIZATION SPECIFIC COMPETENCIES

A facility that handles hazardous materials should have a process safety management program in place. Managing process safety requires individuals with the appropriate process safety competencies. Chapters 2 and 3 defined process safety competency needs and described how to ensure these needs are assigned to appropriate individuals. The Process Safety Competency Matrix (PSCM) summarizes the process safety tasks as defined in the CCPS RBPS Guidelines (CCPS 2007).

Depending on the size of the organization, process safety competencies range from being located at a single facility to being distributed throughout the organization, such as other operating locations, regional or business unit offices, corporate offices, engineering and research centers. In addition outside consultants or contractors may be used to satisfy required competencies. These competencies that may be located outside a specific operating facility would typically be individuals with "Proficiency" levels at the "Expert", or "Leader" levels. For example, some companies may have the company audit function leader in the corporate office as well as the entire audit team ("Experts" and "Practitioners"). In other companies, the audit team members may be taken from other facility sites, or may be supported by contractors. Thus, when determining the process safety competency needs at an individual site, the needs that are being satisfied by individuals at other locations or by contractors must be determined. The remaining competencies should then be filled by individual's located onsite.

Although the practical process safety competencies ("Basic Knowledge", "Practitioner", "Expert" and "Leader") may be defined in a large corporation, the need for awareness throughout the organization, especially at the corporate level is essential.

Management of process safety has to have broad support from the top levels of the corporation in order to receive the necessary priority and resources to be successful. CCPS has identified this need and developed resources to convince top management that process safety is critical to the business. In 2003, CCPS published *The Business Case for Process Safety* and updated the document in 2006 (CCPS 2006). This publication defined four benefits of process safety:

- Corporate responsibility
- Business flexibility
- Risk reduction
- Sustained value

In 2013, CCPS published a video and slide presentation: *Inspiring Process Safety Leadership: The Executive Role* (CCPS 2013c). The purpose of the publications is to "Raise awareness and encourage a dialogue on process safety, as well as process safety culture, at the senior most levels in the company."

When process safety competencies are met by individuals who are not located onsite, it is important to ensure that these individuals will be available when needed. For example, if a mechanical integrity expert in corrosion is located at an engineering location, that individual would need to be available to the site in case of a failure caused by corrosion or an impending failure identified through testing or inspection. Where specific competencies may need to be accessed quickly (say within 24 hours of an identified need), a pool of resources or defined alternates should be identified.

4.2   ASSURE COMPLIANCE WITH REGULATIONS

Some process safety competencies may be required to meet regulatory requirements, industry standards or specific company requirements.

In the US, the OSHA Process Safety Management regulation (29 CFR 1910.119) and the EPA Chemical Accident Prevention Provisions regulation (Title 40 Part 68) define specific process safety competencies. For example, OSHA PSM requires that PHAs be conducted using one of six methodologies or an appropriate equivalent methodology (See OSHA PSM regulation 1910.119(e)(2)(vii)) :

1. What-If
2. Checklist
3. What-If/Checklist
4. Hazard and Operability Study
5. Failure Modes and Effects Analysis
6. Fault Tree Analysis

OSHA PSM also requires that at least one PHA member be knowledgeable in the specific process hazard analysis methodology used. This implies that the PHA leader has an "Expert" level proficiency through experience and/or training. This requirement could be satisfied by a company individual trained in the PHA methodology or an outside consultant. The regulation also specifies that one member of the PHA team has experience and knowledge specific to the process being evaluated. This requirement is generally assumed to mean an operator.

The regulation also requires that the PHA be conducted by a team with expertise in engineering and process operations. These additional competencies may include:

- Process engineering
- Project engineering
- Instrumentation and electrical engineering
- Process technology expertise
- Mechanical integrity expertise

A similar requirement is specified, in OSHA PSM, for incident investigation which states that one person on the investigation team must be knowledgeable in the process involved (again this is generally assumed to be an operator or someone from operations) and other persons with appropriate knowledge and experience to thoroughly investigate and analyze the incident. The other individuals may include a root cause expert and a forensic expert.

In Europe, the Seveso Directive, Directive 96/82/EC as amended (Seveso II), is a prescriptive regulation, in which every clause has competency implications. For example, the facility needs to have a competent accident investigation team that is prepared to address the following once an accident occurs:

- Inform the competent authority of the steps envisaged:
- Alleviate the medium- and long-term effects of the accident, and
- Prevent any recurrence of such an accident.

In the UK, regulations state the requirement to document competencies in safety management systems. The Control of Major Accident Hazards Regulations 1999 (COMAH) and 2005 amendments, which are the enforcing regulations for Seveso within the UK, requires the safety report to document "information to demonstrate to the competent authority that all measures (including competencies) necessary for the prevention and mitigation of major accidents have been taken". COMAH is applicable to establishments storing or handling large quantities hazardous chemicals.

The UK Offshore Installations (Safety Case) Regulations 2005, which seek to reduce the risks from major accident hazards in response to the recommendations of the Cullen Report whose focus was the 1988 Piper Alpha North Sea oil production platform disaster, requires a safety management system (SMS) for managing residual risks. Specifically, the SMS is required to document the "employee selection, competency, training and induction". Norway and Australia have similar regulations.

In 1993, the International Labor Organization released its Prevention of Major Industrial Accidents Convention (C174) in order to prevent major accidents. Seven years later, to fight against major accidents, the Chinese national standard on Identification of Major Hazard Installations (GB 18218-2009) was released in 2000 and then amended in 2009. It was not till 2010, that the State Administration of Work Safety (SAWS) released its first Process Safety Management regulation in China (see *Chemical industry process safety management implementation guide* (AQ/T 3034-2010)*).* This Chinese standard adopted twelve of the fourteen OSHA PSM elements, except for Trade Secrets and Employee Participation, and came into effect on May 1, 2011.

There are numerous recognized and generally accepted good engineering practices (RAGAGEP) that may apply to process safety. The OSHA PSM regulation contains two references to RAGAGEP. The first is in the Process Safety Information element and states: "The employer shall document that equipment complies with recognized and generally accepted good engineering practices (RAGAGEP)." The second reference is in the Mechanical Integrity element and states: "Inspection and testing procedures shall follow recognized and generally accepted good engineering practices."

The CCPS RBPS Guidelines define a specific element to address these requirements: Compliance with Standards. Within these standards, there may be specific process safety competencies defined for various activities such as welding, pressure vessel inspection, corrosion prevention and control, and implementation of nondestructive testing techniques (see *CCPS Guidelines for Mechanical Integrity Systems* (CCPS 2006a)). These competencies can be achieved through training and certifications by outside agencies.

Company standards may require specific process safety competencies that need to be included. For example, a process safety compliance audit team leader may be required to be certified by an outside organization, such as the Certified Process Safety Auditor accreditation provided by BEAC (The Board of Environmental, Health, and Safety Auditor Certifications). Another example might be the Professional Process Safety Engineer Standard from IChemE, with which the organization's process safety expert would meet many of the required competencies for their position. Refer to Appendix 1 for example of process safety auditing competency.

4.3 EXAMPLE TEMPLATES AND CHECKLISTS

The PSCM provided in these Guidelines is a consensus of the subcommittee members. In this section we will provide some example templates and checklists used by other companies to address their process safety competency needs.

4.3.1 Process Hazards Management Coordinator and Hazard Assessment Facilitator

This document (see Appendix 2) provides the process safety competency requirements for the position of Process Hazards Management (PHM) Coordinator and Hazard Assessment Facilitator in a large international corporation.

The first table in the document (Table A2.1 *Hazard Levels Definitions*) defines which processes within the company require a PHM program based on the quantity and hazard of the chemicals being handled. Five hazard levels (I-V) have been defined, of which four require a PHM program.

The second table (Table A2.2 *Facility Level Assignment*) defines three facility safety levels based on risk.

The third table (Table A2.3 *Facility PHM Requirements Based on Hazard Level*) provides the required competencies for both the PHM Coordinator and Hazard Assessment Facilitator based on the hazard level. In this example, the process safety competency requirements are based on the hazard level of the process being managed or reviewed.

The final table (Table A2.4 *Facility PHM Competency Requirements Based on Hazard Level*) then provides the experience and training requirements to meet the required process safety competency needs.

4.3.2 HAZOP Facilitator

Appendix 3 contains an example of another process safety competency matrix for a number of process safety positions at both the facility and corporate level.

Five proficiency levels have been defined:

- Awareness
- Basic application
- Skillful application
- Mastery
- Expert

The above proficiency levels for HAZOP facilitator, match closely with the PSCM proficiency levels (e.g., "Mastery" matches up with "Leader").

The second document within Appendix 2 is a Curriculum Map for the skill of hazard and operability (HAZOP) facilitator. This map contains the technical level descriptors for each proficiency level as well as the training and experience requirements.

4.4   REFERENCES

CCPS 2006. Center for Chemical Process Safety, *The Business Case for Process Safety,* Second Edition, New York 2006.

CCPS 2013c. Center for Chemical Process Safety, Video and PowerPoint Presentation: *Inspiring Process Safety Leadership: The Executive Role,* New York, 2013.

CCPS 2007. Center for Chemical Process Safety, *Guidelines for Risk Based Process Safety*, New York, 2007.

CCPS 2013a. Center for Chemical Process Safety, Guidelines for Managing Process Safety Risks During Organizational Change, New York, 2013.

CCPS 2006a. Center for Chemical Process Safety, Guidelines *for Mechanical Integrity Systems,* New York, 2006.

29 CFR 1910.119, *Process Safety Management of Highly Hazardous Chemicals: Explosives and Blasting Agents; Final Rule*, US Department of Labor, Occupational Safety and Health Administration, 24 February, 1992.

Title 40 Part 68, *Chemical Accident Prevention Provisions*, US Environmental Protection Agency, January 31, 1994.

Directive 96/82/EC, on the control of major-accident hazards involving dangerous substances, The Council of the European Union, 9 December 1996.

COMAH. Control of Major Accident Hazards Regulations 1999 and 2005 amendments, UK Health and Safety Executive, 30 January 2005.

Offshore Installations (Safety Case) Regulations 2005, UK Health and Safety Executive, 9 November 2005.

http://www.aiche.org/ccps/resources/education/process-safety-leadership/executive-role

C174, Prevention of Major Industrial Accidents Convention, 1993 (No. 174)

GB 18218-2009, *Identification of major hazard installations for dangerous chemicals*, National Standard of the People's Republic of China, 31 March, 2009.

AQ/T 3034-2010, *Chemical industry process safety management implementation guide,* National Standard of the People's Republic of China, 6 September, 2010.

5

ASSESS COMPETENCIES VS. NEEDS

A competency Matrix assessment provides a baseline of the organization's existing process safety competencies and helps to evaluate which job roles currently have adequate process safety skills, knowledge, and expertise. The initial assessment can be compared to the organization-specific competency Matrix that was customized in Chapters 3 and 4, to determine the remaining competencies that are not already fulfilled.

5.1 ASSESSING EXISTING COMPETENCIES

The Matrix can be used as an assessment tool to identify existing process safety competencies within an organization. Each employee within the organization can perform an individual self-assessment to document the skills and knowledge he or she possesses. Managers and peers can also conduct assessments of individuals performing the role, to determine the competency gaps between the individuals' experience and expertise, and those required for the role being considered. The management assessment will be able to identify if the individual is qualified and possesses the right skills and knowledge for the specific role.

The "Skills & Knowledge" worksheet provides a base template for a competency assessment tool. The worksheet can be converted into an assessment tool with a simple highlighting method to display which competencies in the Matrix have been fulfilled. This can be done using many Microsoft ® Excel features, formatting, or functions, based on the user's preference. To minimize the need to significantly modify the template worksheet, the user can use cell color to highlight competencies, and with newer versions of Microsoft ® Excel, the Data Sort or Filter functions can sort or filter by cell color.

5.1.1 Self-Assessment

For an individual self-assessment, the assessor would review the Skills and Knowledge worksheet element by element and identify the competencies that he or she currently meets. The assessor would then document his or her proficiency level based on whether he or she identified all the level-specific competencies listed for an element. Once a proficiency level for each element has been determined, the assessor would review the definitions in the "Proficiency Level" worksheet and comment about his or her appropriate proficiency level in the assessment comments column in the "RBPS Table" worksheet.

5.1.2 Peer / Manager Assessment

For a peer or manager assessment, the assessor would review the skills and knowledge worksheet element to assess the competencies that a certain position should have. The management assessment would typically be conducted by an experienced individual or group within the organization. With the understanding of the existing skills, knowledge, and expertise that a person in that position currently possesses, the assessor would document the expected competencies for that position.

5.2 TRAINING FOR ASSESSORS

It is likely that the sample of employees that are chosen to perform the self and peer / manager assessments will not have been involved in the Matrix customization. It will be necessary to train the assessors in the components of the Matrix that are described in chapters 2 and 3, as well as how to use the assessment tool before starting the assessment.

Those chosen for the assessment group should have the process safety knowledge necessary to recognize the existing process safety competencies. For roles that have multiple employees in the same role, a sample of employees can be chosen for the assessment group. Multiple assessments of the same role will show a range of competency data that can be used to determine an organization-recognized, minimum proficiency level for that role

Training topics for the assessors should include:

- The intent of the competency Matrix and the assessment
- Explanation of the components of the Matrix and RBPS elements, to include customization and organization-specific competencies
- Explanation of the components of the assessment tool

5.3   IDENTIFY GAPS BETWEEN CURRENT STATUS AND NEEDS

The information gained in these assessments gives a current picture of the status of each job role's process safety competency. This information can be used to identify the gaps between the current status and organizational need that was mapped out in the customized Matrix.

A gap analysis can be done for each element on the skills and knowledge worksheet. When reviewing the assessment information with the customized Matrix, a second RBPS table (RBPS2) can be created, showing the current status of each job role. This would give two tables, that when compared side-by-side, can show the gaps and differences.

The analysis can be used to show the following gaps:

- The elements on the RBPS table that do not have someone in the organization filling the need for process safety competency (shown as a blank row)
- Which job roles lack process safety competency (blank column), and
- In which competencies, in the skills and knowledge tab, the job roles need to improve their skill, knowledge, and experience to show competency (proficiency levels on the RBPS2 are lower than the RBPS table).

The analysis can also highlight where an employee may be competent in elements unrelated to his or her job role (proficiency level on the RBPS2 is higher than the RBPS table).

These gaps, when compiled into findings and recommended items, will set the framework to develop the gap closure plan, which is further discussed in the next chapter.

6

DEVELOP GAP CLOSURE PLAN

The assessment and gap analysis from Chapter 5 identifies the gaps between the current status and competency needs. The findings and recommendations from the gap analysis are the basis for developing a gap closure plan.

To draft the gap closure plan, a core team can review the findings and recommendations from the gap analysis. This review will include the team's determination on which method is appropriate to address the gaps, and the development and assigning of action items.

As discussed in Section 5.3 *Identify Gaps Between Current Status And Needs*, the analysis could show any of the following gaps:

- The elements on the RBPS table that do not have someone filling the need for process safety competency
- Which individuals, in their job roles, have lack of process safety competency
- In which competencies, in the Skills and Knowledge tab, the individuals need to improve their skill, knowledge, and experience to show competency in their job roles

To address these gaps, a variety of training and non-training methods can be used, and supporting materials can be developed to supplement the gap closure action items. Additionally, prerequisites can be developed for the competency levels, as stepping stones to be completed before advancing to the next competency level.

6.1 METHODS FOR CLOSING THE GAPS

Where there are competency gaps, the review team will need to discuss the methods to close those gaps and determine which method is most appropriate. Ideally, incumbents in process safety roles will possess all the necessary competencies to fulfill the criteria established by the company. If incumbents do not possess the necessary competencies, there are a number of options available to establish the required competencies.

6.1.1 Tasks or Personnel Reassignment

When a process safety competency gap has been identified and the requisite skill set is available within the company, the person with the desired competency may be reassigned to support the process safety efforts where the gap is identified. This stop-gap approach may be taken until such time that another person within the desired team can develop this competency on a long term basis.

6.1.2 Internal & External Training

To fill a process safety competency, a person may be chosen and trained to the competency of that element. In many instances the training required for some of the lower proficiency levels can be met internally through either formal, such as classroom-based training sessions, or more informal one-on-one training, such as on-the-job supervision, coaching, or mentoring, depending on which option can meet the needs in the most efficient and timely manner.

For the "Expert" proficiency level, specific training courses or certifications may be needed. For example, a PHA facilitator may best receive external training to learn about a specific hazard assessment methodology. It may be best to require external training for competencies that need to be met at a higher proficiency level, such as understanding facility siting, human factors, or complex unique processes.

6.1.3 External Resources

If the competency doesn't exist within the company, hiring someone into the company with the skills, knowledge, or expertise to close that gap is a possible solution. This solution may take a significant amount of time to find a person with the requisite competencies, and hiring is often viewed as a long term commitment on the part of the company.

If the competency is needed only for the short-term, such as to support a specific project phase, then an alternate solution is to work with a consultant or someone available with the needed process safety expertise for certain projects that require that competency. It may be necessary to review how filling a competency on a project-basis affects the entire scope of the competency assessment before choosing this solution. A good exercise is to document the tasks and projects that would rely on this process safety competency before filling a competency on a project-basis.

6.2 SUPPORTING MATERIALS

Since some gaps may show that there is not an internal resource with the specific process safety competency, the team can look to supporting materials and other resources to determine how to close their organization competency gaps. There are many resources which can be accessed to enhance process safety knowledge which can lead to competency development, some are outlined here:

- Many books are available which address a broad range of process safety topics from notable authors such as Trevor Kletz and respected international process safety organizations such as CCPS and IChemE
- Numerous training courses are available, including e-learning, webinar, and classroom training courses

- Technical presentations covering individual elements of process safety and underlying technical and managerial concepts can be found on the CCPS website and are available at recognized process safety conferences, such as the Global Congress on Process Safety
- Guidance documents describing engineering best practices can also provide information to help close gaps. Websites of regulators such as the UK Health and Safety Executive and US OSHA for guidance regarding their process safety regulations
- Standards setting organizations also provide guidance on appropriate process safety controls. Examples include ISO, OECD, or BSI for international application or API, ASME, and NFPA in the United States
- In addition, the detailed incident investigations and associated animations published by IChemE and the US CSB contain many lessons learned related to RBPS elements
- Universities may have process safety related material and studies. A few universities offer a process safety course to supplement their Chemical Engineering curriculum, such as the University of Sheffield in the UK. Texas A&M University has the Mary Kay O'Connor Center for Process Safety which performs process safety research and offers classes in process safety
- Research journal articles and presentation papers are available online or in hardcopy formats
- Consultants can also be hired to recommend how to close gaps with limited resources

All of these resources can be used to guide the team to determine the best course of action to close a gap.

6.3 PRE-REQUISITES BEFORE PROGRESSING TO THE NEXT LEVEL

Appendix 3 (Table A3.2 *HAZOP Proficiency Level Example*) shows pre-requisites examples for someone meeting the HAZOP element at the "Awareness" and "Basic" levels. For an employee to meet the "Awareness" level, they need to attend the internal training sessions for new employees and complete a course on the Fundamentals of Process Safety. Before meeting proficiency at the "Basic" level, the employee needs to meet the pre-requisite and take the General Hazard Recognition and Process Hazard Recognition internal training.

Pre-requisites can be defined in this example to ensure additional skills are achieved, outside trainings taken, or other knowledge is gained prior to advancing to the next proficiency level. Before becoming a hazard evaluation facilitator (Proficiency Level of "Skillful Application" or higher), the candidate can be expected first, to have been a hazard evaluation team member for a number of on-site hazard evaluation teams. This can also apply to other leader roles, such as incident investigations team leader, emergency coordinator or audit team leader. Defining pre-requisites should help ensure consistent competency across the organization and for a job role to fill a competency.

6.4   EXAMPLE OF MANAGING GAP CLOSURE

Appendix 4 addresses how process safety competency progress can be shown on the PSCM using the percent complete scale. With a gap closure plan established, the gaps in the Matrix can be used to track progress.

7

SUSTAINING COMPETENCIES

7.1 STRATEGIES FOR SUSTAINING COMPETENCIES

With a fully-customized competency Matrix, the last step is to create a management system to ensure that it is maintained up-to-date, so that as a result of organizational changes, competencies are not lost or overlooked. A document control system for the Matrix may include assigning ownership of the document and establishing a frequency for review and revision, to ensure that changes in competency needs are properly documented.

7.1.1 Apply Changing Technology & Lessons Learned

Other activities can highlight the need to re-assess and update the competency Matrix. Incident investigation, audit findings, or a formal management review, can alert the organization to a process safety gap or lack of competency that needs to be addressed. Lessons learned that come out of these activities can be included in a competency Matrix review and gap closure plan.

When required competencies change, such as when a process safety management system is changed, an assessment may be performed to determine if additional resources are required. This would trigger the need for a competency Matrix review and revision.

Competencies would change when outsourced process safety services are in sourced, or when internal process safety resources are overloaded and the decision is made to work with process safety consultants.

As for individuals maintaining their own process safety competency, a variety of topics are discussed at professional conferences. The process safety industry evolves with the changes in technology and learnings from incidents and operational process safety implementation. Conferences such as the AIChE/CCPS Global Congress on Process Safety, AIChE/CCPS Latin American Process Safety Conference, AIChE/CCPS Asia-Pacific Process Safety Conference, AIChE Ammonia Safety Conference, EFCE Loss Prevention Symposium, IChemE Hazards Symposium and API Center for Offshore Safety Annual Forum can be attended to sustain and develop process safety competency. Often attendance of such events is seen as demonstrating continued professional development by institutions such as the AIChE and IChemE.

7.1.2 Metrics

The application of metrics to the system to sustain process safety competency will give the ability to measure and improve performance relative to the competency Matrix.

The following list provides examples of some metrics that could be assessed when the Matrix is reviewed (CCPS 2011):

- % of competencies not fulfilled organization-wide
- % of competencies filled externally
- % of incident investigations that identify a lack of competency as an incident cause
- % of refresher training completed (process safety awareness or any prerequisite)
- % of action items open from previous gap analysis

As an organization becomes more competent in process safety, the culture of the organization should begin to change. To measure organizational process safety culture, a culture survey could be given, with the assurance that it is documented anonymously. As the process safety competencies improve and change throughout the organization, a re-administration of the survey gives data that can be measured. Example culture surveys which address process safety competencies, such as included in the CCPS Vision 2020 project assessment, the Baker Panel culture survey, or the OECD Corporate Governance for Process Safety, Guidance for Senior Leaders in High Hazard Industries can be used as models to develop company-specific culture survey.

7.1.3 Demonstrate and Verify Competencies are in Place and Effective

The competency Matrix may be reviewed during an audit or incident investigation. These reviews can help determine if the competencies are in place and effective. Implementation verification can be done through interviews, document review, and on an informal basis through on-the-job refresher training and testing.

7.2 REVIEW AND UPDATE COMPETENCY NEEDS

Managing process safety competency during organizational change will highlight the need to update individual competencies when an organization is dealing with the replacement of a subject matter expert or any position that directly effects process safety. The already established process safety competencies form the basis for ensuring that these competencies are maintained as changes are made in the organization (see CCPS *Guidelines for Managing Process Safety Risks During Organizational Change* (CCPS 2013)). A gap analysis can be performed to determine the gaps that may have formed with the organizational change. Chapter 3 discusses how to use and update the Matrix during an organizational change.

7.3 ORGANIZATIONAL PROCESS SAFETY CULTURE

A healthy and sustained organizational process safety culture comes from the visible leadership of executive management. This confirms the consistent vision of process safety that can be adapted to the current state of the organization.

Just as the organization needs a strong process safety culture, so it needs strong operational discipline. Operational discipline is necessary to ensure that operations and processes are being performed accurately and safely every time (see CCPS *Conduct of Operations and Operational Discipline: For Improving Process Safety in Industry* (CCPS 2011a)). "Every site has unique operational-discipline-related strengths and weaknesses[1]." With this understanding of the organization-specific characteristics, the training plan can be customized to reward the strengths and support the weaknesses, and sustain competent performance.

7.4 REFERENCES

CCPS 2011. Center for Chemical Process Safety, Process Safety Leading and Lagging Metrics: You Don't Improve What You Don't Measure, New York, 2011.

Corporate Governance for Process Safety, Guidance for Senior Leaders in High Hazard Industries, OECD Environment, Health and Safety, Chemical Accidents Programme, June 2012.

CCPS 2013. Center for Chemical Process Safety, Guidelines for Managing Process Safety Risks During Organizational Change, New York, 2013.

CCPS 2011a. Center for Chemical Process Safety, Conduct of Operations and Operational Discipline: For Improving Process Safety in Industry, New York, 2011.

---

[1] Klein, J.A. and Vaughen, B.K., "Implement an Operational Discipline Program to Improve Plant Process Safety", *Chemical Engineering Progress,* New York, 2011.

APPENDIX 1: EXAMPLE COMPETENCIES FOR AUDITING

Table A1.1 is a list of items relevant to competency (knowledge, skills, and experience) that can be used for demonstrating auditing competence:

Table A1.1.   Demonstration of Competency

Formal Assessment	Occupational Training	Previous Employment	Certificate / License	Education
• Observed performing activity safely, proficiently • Demonstrated skills at worksite • Demonstrated skills by simulation • Oral questioning • Passed written knowledge assessment on activity • Peer review • Information compiled by employee	• Internal training • External training • Vendor specific equipment training • Equipment or process specific e-learning • Formal on-the-job training	• Performing audits in field conditions • Performing audits with another company • Receiving mentoring or coaching while performing audits • Mentoring, coaching or training others to perform audits • Performance review • Performance record(s) • Drills • Management observes performance - routine • Procedural review • P & ID / facility flow diagram review • Internet review of equipment properties / specifications • Informal on-the-job training • Observed others performing tasks • Reading from assigned list • Completed independent e-learning course(s)	• Certified for this position per specific standard • Licensed for this position per specific standard	• High school diploma or equivalent • Vocational training • Apprentice program • College / University • Advanced degree(s)

APPENDIX 2: PHM COORDINATOR & HA FACILITATOR QUALIFICATIONS

A2.1 PURPOSE

Provide resource qualification guidance and describe performance expectations of the:

- Process Hazard Management (PHM) Coordinator
  - o Manages the overall PHM program of the facility
- Hazard Assessment (HA) Facilitator
  - o Plans, facilitates and leads required hazard assessment sessions of covered processes

A2.2 ASSUMPTIONS

- Process Hazard Management is a program. It does not replace any regulatory requirements or industry standards.
- PHM is a Process Safety program designed to manage risk to an acceptable level of potentially catastrophic severity level hazards (Risk Assessment Ranking events that score above 1600).
- Hazardous Process Evaluation Tool (HazPET) assessment, defined chemical process hazard level I thru V definitions apply PHM elements as per PHM Guidance manual.
- Full ownership of the PHM program execution for a covered process should reside with the process owner at the facility and not the PHM coordinator.

A2.2.1 HazPET level & EHS Safety Level

The Hazardous Process Evaluation Tool (HazPET) level represents the hazard level of a facility based on the chemical(s) physical properties and quantities. This is a hazard level, not a risk level, since facility controls are not considered in the analysis.

A facility with more than one physical chemical process should assign a hazard level to each physical process. The chemical process is defined by the physical process equipment as described by the process flow, and not solely by its product. When the same general process uses different chemicals for multiple products, the worst credible case should be used.

Table A2.1.   **Hazard Levels Definitions**

HazPET Level	Hazard Level I	Hazard Level II	Hazard Level III	Hazard Level IV	Hazard Level V
Total PHM Decision Flow	- < 100lb Hazardous Chemicals. -No NFPA 3 (500lbs) or 4 (250lbs). -Not on Hazardous process List. -< 25% TQ	> 100lb Hazardous Chemicals. Less than TQ. -Has NFPA 3 (500lbs) or 4 (250lbs) or > 25% TQ and determined by Division, Regional, or Corporate Safety hazard review.	- > 100lb Hazardous Chemicals. Less than TQ. -On the Hazardous Process List	-Listed Chemical(s) exceed TQ. -On the Hazardous Process List	-Listed Chemical(s) exceed TQ & process run under extreme conditions or characteristics. On the Hazardous Process List.
HazPET	Score of <45	Score of 45 - 100	Score of 101 - 200	Score of 201 - 400	Score of >400
PHM Requirements	Not Required	Required	Required	Required	Required

Reference: PHM Guidance Manual

Table A2.2.   Facility Level Assignment

Score Card Safety Facility Level Alignment	Facility Safety Level 1	Facility Safety Level 2	Facility Safety Level 3
Level of Process Risk at Facility (chemicals handled, complexity of equipment)	Low risk of spill, fire, explosion or injury Basic / well known processes No PSM covered processes Incidental chemical use	Moderate risk of spill, explosion, fire or injury Varied, complex or unique processes Few PSM covered processes	Significant risk of spill, explosion, fire or injury with possible off-site impacts Many varied, complex or unique processes Many PSM covered processes
General Description	Has simple needs or few activities, with activities being relatively low risk to employee health and safety and to the environment.	Moderate number of needs and/or moderately complex program.	Has complex needs and a large number of activities, with potentially more significant risks to employee.

Reference: EHS Directory

A2.3 FULL TIME EQUIVALENT (FTE) RESOURCE ALIGNMENT

FTE resources is an organizational decision based on the number of PHM processes, level of hazard, assessment of risk, needs, complexity, structure, geography, etc. Due to these many variables there is not a set formula for allocation. Full and part FTE justification is determined by the facility in coordination with parent organizational levels.

PHM can be supported with resources from an adjacent and, or, higher levels of the organization.

Full ownership of the PHM program execution for a covered process should reside with the process owner at the facility and not the PHM coordinator.

A2.3.1 Process Hazard Management Coordinator and Hazard Assessment Facilitator Qualification

A2.4 EXPERTISE AND EXPERIENCE

The chart below uses the HazPET hazard levels coupled with the alignment of facility safety levels to provide guidance for the PHM Coordinator and Hazard Assessment Facilitator expertise and experience. Interpretation between Coordinator and Facilitator expertise and experience may be required based on position assignments. A large majority of personnel are assigned both duties.

- Facilities with higher hazard levels and process complexity require a higher level of experience and expertise. This should be determined at the facility level with guidance from higher levels in the organization.
- This chart provides a tiered description of the PHM Coordinator and Hazard Assessment Facilitator qualifications based on facility hazard levels. Each description builds on the one below.

A2.5 EHS AND PHM ALIGNMENT

- PHM Coordinator and EHS management responsibilities in many program elements overlap requiring a clear understanding of ownership and collaboration within the facility.
- PHM functions do not require the qualifications of the EHS function.
- In many industries PHM functions under the engineering department instead of EHS. These responsibilities fit well within engineering since the bases of PHM are process design, reliability, maintenance, and operations.
- Placing PHM in the engineering function may offer a more flexible engineering career track and improve resource

supply by tracking the PHM function outside the EHS career field.

- This does not relieve EHS managers from oversight, regulatory or other assigned PHM responsibilities.

Table A2.3.   Facility PHM Requirements Based on Hazard Level

	Facility PHM Level 1	Facility PHM Level 2	Facility PHM Level 3	
	Hazard Level I	Hazard Level II	Hazard Level III	Hazard Level IV	Hazard Level V

Advanced and specific experience with Chemical Processes to include:
- Quantitative Failure Analysis
- Experience with High pressure and exothermic controls

Direct experience with Chemical Processes to include:
- Experience in design, controls, operations, reliability, and maintenance principles as related to the process
- Failure analysis

General experience with Chemical Processes to include:
- Understand process control, reliability, maintenance and human interface principles,
- Prevention & mitigation technique
- Process chemistry

PHM Foundation:
- Engineering, Technical, or EHS higher educational degree and / or adequate experience related to the process hazards and manufacturing methods.
- Completed PHA Facilitator training and have a solid understanding and ability to execution all PHM program elements ….to include all foundation PHA facilitator qualifications
- General techniques of program / project management / communication skills
- A working knowledge and application of risk management principles, basic chemical process controls, and safety controls.
- Qualitative Hazard Analysis & Risk Assessment
- Understand requirements for the control and mitigation of toxic and flammable chemicals to include explosive

PHM Not Required

A2.6 OVERVIEW OF DUTIES AND RESPONSIBILITIES

A2.6.1 PHM Coordinator

- Responsible for the performance, quality and improvement of the PHM program and its objective of preventing a facility catastrophic event
- Program management
    - Continuous improvement
    - Trending, evaluation, measurement
    - Self-assessment, metrics
    - Planning and implementation
- Teach / train, as required, aspects of PHM
- Develop, utilize, as needed, procedures, tools, and applications
- Communicate and coordinate risk information, needs, expectations, status in all directions of the organization
- Assist process owners in execution of PHM elements and program
- Provide necessary skills in terms of hazard assessment facilitation, documentation, and required resources for a successful hazard assessment
- Participation in PHM steering team at the location.
- Work closely with EHS management in the coordination of applicable PHM elements

A2.6.2 Hazard Assessment Facilitator

- Responsible for the management of hazard assessments, revalidations process, and redos (PHA revalidation starting with blank worksheets)
- Responsible for the technical execution of hazard assessments, redos, and revalidations

- Provides hazard assessment training to the selected hazard assessment team
- Communicates results as required

NOTE: PHM coordinators and facilitators will be listed in the EHS Directory.

A2.7 COMPETENCY-BASED KNOWLEDGE (TRAINING) ROAD MAP FOR QUALIFICATION

This training road map focuses on subject matter topics that are intended as a minimum guideline for areas of expertise based on the hazard level and process complexity. Each facility should complete an evaluation and analysis that identifies specific needs for training and qualifications. For additional assistance in PHM resource allocation and planning contact the Corporate Safety group.

Table A2.4.   Facility PHM Competency Requirements Based on Hazard Level

Facility PHM Level 1	Facility PHM Level 2		Facility PHM Level 3	
Hazard Level I	Hazard Level II	Hazard Level III	Hazard Level IV	Hazard Level V
				• Required: Training in LOPA, SIS, Quantitative Analysis • Recommended: Training as needed by process complexity
			• Required: Training in mechanical Integrity methods, Failure analysis (vendor or organization provided) • Required: Training as needed by process complexity (vendor or organization provided) • Recommended: Advanced training; Facility Siting, Human Factors, Advanced hazard assessment, (vendor)	
		• Required: Training in process control, prevention & mitigation techniques for flammables and hazardous dust (Facility, Division, Region, or Country provided) • Required: Training on the facility process chemistry (Facility, Division, Region, or Country provided) • Required: Overview of basic regulatory and engineering standards (Facility, Division, Region, Country provided)		
PHM Not Required	• Recommended for selection: Engineering, Technical, or EHS higher educational degree and / or adequate experience related to the process hazards and manufacturing methods. • Required: Hazard Assessment Facilitator course (Includes PHM Coordinator training) (instructed by PHM subject matter expert as designated by Corporate PHM Group) • Required: Prior to independent facilitation of a hazard assessment, or revalidation, the facilitator will be coached by an experienced hazard assessment facilitator for a minimum of 2 analyses. (Facility, Division, Region, Country provided) • Recommended: Overview of Risk Assessment workbook (Facility, Division, Region, or Country provided) • Recommended : Training in Qualitative Hazard Analysis, Risk Assessment, Risk Management principles (Vendor or organization provided)			

APPENDIX 3: HAZOP FACILITATOR

Table A3.1 is an example of a summary table defining required specific activities linked to the skill "Hazard & Operability Analysis (HAZOP)"; and, table A3.2 is an example of how technical skills, training, experience, and pre-requisites can be assigned to proficiency levels.

Table A3.1.   **Process Safety Element: Process Hazard Analysis (PHA)**

Skill Name	Skill Description
Hazard & Operability Analysis (HAZOP)	Understand and apply HAZOP techniques to identify hazards (what can go wrong), consequences (how bad could it be), and document actions
Specific activities included in this skill	

• Define the scope and objectives of a PHA using the HAZOP technique	• Conduct (lead or participate) the review
• Gather and prepare process safety information for the review	• Conduct supplemental review(s)
• Define the team and skills required	• Document review and action items
• Establish process nodes and depth of review	• Present results
• Define Guide Words, Process Parameters (Variables), Deviations	

Table A3.2.　HAZOP Proficiency Level Example

Awareness	Basic Application	Skillful Application	Mastery	Expert
Technical Level Descriptors:				
Can describe the objectives of HAZOP and when it is used	Ibid as Awareness: Understands the technical aspects and the process safety information of the process being analyzed	Ibid as Basic Application: Can plan the scope and lead a HAZOP review	Ibid of Skillful Application: Defines and documents the HAZOP management system	Ibid as Mastery: Verify the HAZOP practices are effective
Can describe how the Hazard Analysis process relates to process safety	identification techniques and requirements of OSHA PSM and EPA RMP. Alternatively: Understands COMAH / SEVESO PHA	Can identify the required technical skills of team members. Knowledge of corporate standard #XYZ.	Audit other HAZOP studies Provide improvements to HAZOP methods	Teach HAZOP at senior management level and develop learning packages
Can describe the general classes of process hazards	Perform as a team member to use process guide words, process parameters, and deviations	Sets the HAZOP guide word and parameter definitions	Lead HAZOP studies for major processes/ projects	Lead discussions with Regulators, Government Agencies, Contractors and others in the industry to change approach to HAZOP
Know where to find a list of process hazards in their work area		Ensures HAZOP review is completed per its scope, documented, and action items prepared	Business Unit Authority in some areas	Technical authority

Table A3.2   Cont. HAZOP Proficiency Level Example

	Awareness	Basic Application	Skillful Application	Mastery	Expert
Required Training:					
Initial (Necessary to qualify for the position)	New employee process safety training class	Technical training relative to the process area being studied	PHA (HAZOP and What-If) Training	Advanced FHA Training	N/A
Refresher (Any required follow-up or periodic training)		Review all previous HAZOP studies just prior to the next HAZOP	Hazard Identification Training- every 5 years	N/A	N/A
Verification (Any demonstration of competency)		Test covering PHA Process	HAZOP audit results	Technical Authority recognition	Industry wide recognition Publications
Material (Texts, articles, documents, standards or procedures)	Work area process safety information inventory	29 CFR 1910.119 (e)(1-3)Work area specific Process Safety Operating Standards (PSOS) CCPS: "Guidelines for Risk Based Process Safety – chapter 9"	CCPS: "Guidelines for Hazard Evaluation Procedures, 3rd edition" CCPS: "Guidelines for Safe Process Operations and Maintenance"	N/A	N/A

Table A3.2   Cont. HAZOP Proficiency Level Example

	Awareness	Basic Application	Skillful Application	Mastery	Expert
Required Experience:					
Achievements or tenure		Involvement in two or more HAZOP studies	Lead two or more small HAZOP studies	5 years experience in leading major HAZOP process/project studies	AIChE/CCPS, API or similar professional society active participation
Prerequisite					
(Requisite or additional skills also needed)	New employee general orientation Ibid as #2 Fundamentals of Process Safety	Ibid as #3 General Hazard Recognition Ibid as #15 Process Hazard Recognition Can identify hazards			

APPENDIX 4: SHOWING GAP CLOSURE PROGRESS

% Complete	
0%	NO PROGRESS / GAP RECOGNIZED
25%	GAP CLOSURE PLAN DRAFTED
50%	PLAN REVIEWED
75%	PLAN APPROVED / RESOURCED
100%	PLAN IMPLEMENTED / COMPETENCY MET

- Define progressive completion estimates as part of the gap closure plan.
- Estimate deliverable completion progress as percent complete and include in a new row

Table A4.1. Gap Closure Progress

Pillar	Element	Responsibility Assigned To	Operations	Inspection, Testing & Maintenance	Health Safety & Environmental		Corporate or Contractor	Comments
			Engineer	Engineer	Engineer	Manager		
I. Commit to Process Safety	Compliance with Standards	HSE Manager	3	3	3	3	4	An optional expert may be called upon as needed to establish or improve performance in this element.
	Progress Complete	**100%**	0%	25%	100%	100%	100%	

INDEX

A

ASSESSORS 52

C

CCPS Vision 20/20................................. 9
COMAH 44
competency assessment tool 51

E

Engineer.. 17

F

Facility Manager................................. 14

G

gap analysis.. 53
gap closure plan.................................. 55

H

Hazard Assessment Facilitator............ 46
HAZOP Facilitator................................ 47
HSE Manager 14, 15

I

Inspection, Testing, and Maintenance
    Manager.................................. 14, 15

L

Lessons Learned.................................. 61

M

manager assessment 52

Metrics.. 62

O

Operations Manager..................... 14, 15
Operator 16
organization specific competencies.... 41
organizational change......................... 37
Organizational Change Management.. 9
OSHA PSM.. 43

P

pre-requisites examples 59
Process Hazards Management
    Coordinator 46
process safety competencies.............. 43
Process safety competency 5
Process Safety Competency Matrix 1, 23
Process Safety Culture 7, 64
process safety knowledge.................... 8
Proficiency Level worksheet 24, 25
PROFICIENCY LEVELS 18
    Awareness 18
    Basic Knowledge............................ 19
    Expert 20
    Leader.. 21
    Practitioner.................................... 20
Project Engineer 17
Project Manager 17

R

RAGAGEP 45
Risk Based Process Safety 4
    worksheet 25

S

Self-Assessment 52
Seveso Directive 44

Skills & Knowledge.................. **21**, 34, 51
Skills and Knowledge worksheet... **24**, 30
Supervisor 14, 16

T

Technician 14, 16
training 56
training matrix 31